AIRCRAFT

The Supersonic X-15 and High-Tech NASA Aircraft

Henry M. Holden

Enslow Publishers, Inc.

40 Industrial Road   PO Box 38
Box 398   Aldershot
Berkeley Heights, NJ 07922   Hants GU12 6BP
USA   UK

http://www.enslow.com

To Stephen and Scott, my pride and joy.

Library of Congress Cataloging-in-Publication Data

Holden, Henry M.
    The Supersonic X-15 and High-Tech NASA Aircraft / Henry M. Holden.
        p. cm.
    Includes bibliographical references and index.
    ISBN 0-7660-1717-6
    1. X-15 (Rocket aircraft)—Juvenile literature. 2. Space shuttles—Juvenile literature. 3. Space stations—Juvenile literature. 4. Research aircraft—Juvenile literature. 5. United States. National Aeronautics and Space Administration—Juvenile literature. I. Title.
TL789.8.U6 X5155 2002
629.1'07'2073—dc21   2001004122

Printed in the United States of America

10 9 8 7 6 5 4 3 2 1

To Our Readers: We have done our best to make sure all Internet Addresses in this book were active and appropriate when we went to press. However, the author and the publisher have no control over and assume no liability for the material available on those Internet sites or on other Web sites they may link to. Any comments or suggestions can be sent by e-mail to comments@enslow.com or to the address on the back cover.

Photo Credits: © Corel Corporation, pp. 3, 13, 20, 26, 34; Henry M. Holden, pp. 21, 37, 41 (bottom); NASA, pp. 4–5, 7, 8, 12, 15, 17, 18, 19, 22, 23, 25, 27, 28, 29, 31, 32, 36, 38, 39, 40, 41 (top), 42.

Cover Photo: NASA

Contents

CHAPTER 1

Flying the Supersonic X-15

20003

X-15 attached to wing of B-52

"**T**en seconds to drop," the voice over Bill Dana's earphones said. It was November 4, 1965, and Dana was in the black needle-shaped X-15 rocket plane, 45,000 feet above Earth. Dana, a test pilot, was trying out this new rocket plane. He was about to be air-launched in the X-15, which was attached to the wing of an eight-engine B-52 bomber.

"Five, four, three, two, one—release," said a voice in the NASA Flight Research Center control room at Edwards Air Force Base in California.[1] Almost immediately after the X-15 dropped from the B-52, its rocket engine fired. A single white contrail streaked across the blue sky. Contrails, or condensation trails, are steam trails formed by the hot engine exhaust. The contrail quickly pulled ahead of the B-52. Leaving the wing of the mother ship felt like being fired off by some hidden cannon.[2] Within seconds, the X-15 accelerated well beyond the B-52.

The National Aeronautics and Space Administration (NASA) built three high-speed airplanes called the X-15. They flew from June 1959 to October 1968. They were used to explore the space surrounding Earth and to study the effects of gravity, high speed, and high altitude on humans as well as their effects on the plane.

The X-15 was one of the most powerful research airplanes built. To minimize the fuel burned on takeoff, the X-15 was dropped from the wing of a mother ship. In Dana's case, the mother ship was a B-52. The X-15 carried only enough rocket fuel for about two minutes. For the remainder of its flight, about eight minutes, it was without power.

The rocket engine of the X-15 produced over 600,000 horsepower. For comparison, a car engine produces about 150 to 200 horsepower. Engines need oxygen to burn fuel, but there is no oxygen in space. Therefore, the X-15

Seconds after it fires its rocket engines, the X-15 appears to be streaking toward the Sun. Its contrail is the only way to follow this rocket plane.

carried about 1,000 gallons of liquid oxygen and 1,445 gallons of anhydrous ammonia gas. When these two chemicals came together, they created the propellant for the engine to work in space.

≡ *Flying in Space*

When the engine of the X-15 shut down, Dana was in microgravity. He felt nearly weightless. Gravity is what holds things to the ground. Microgravity describes a very weak effect of gravity. Microgravity gave Dana relief from the multiple g-force acceleration that he felt upon takeoff.

Bill Dana was one of the pilots who flew the X-15. On the top of the plane's nose and on each side are four of the eight small hydrogen peroxide thrust rocket nozzles used to control the X-15 at very high altitudes.

During the acceleration, he went from a subsonic speed to nearly six times the speed of sound (about 4,500 miles per hour). The g forces pressing him into the seat were equivalent to up to four times his body weight.

Flying in space is different from flying near Earth's surface. There is almost no air in the high reaches of the atmosphere where the X-15 flew. But a plane's surfaces need air to flow over them so that the plane can change direction. This design would not work in space.

The X-15 used a new method to steer the aircraft. There were small hydrogen peroxide thrust rockets located on the nose and wings of the plane. These rockets steered the X-15 at high altitudes. To make the airplane move to the left, the pilot pushed the control stick to the left. Two small rockets fired on the right side of the nose and pushed the aircraft to the left. To turn right, the pilot moved the control stick to the right to fire the rockets on the left side of the nose. Additional thrusters on the wingtips controlled the roll.

To control the pitch, or up-and-down motion, a pilot pulled back or pushed down on the control stick. Pulling back on the control stick fired two small rockets on the bottom of the nose, forcing the nose up. Pushing the control stick down fired a small rocket on the top of the nose, forcing the nose down. Today's space shuttles use a similar steering system when they are in space. At lower altitudes the X-15 pilot used the horizontal and vertical control surfaces on the tail of the airplane to control its direction.

After the eight- to ten-minute flight, the X-15 returned to Earth. Since it had no fuel remaining, it glided to a landing at about 210 miles per hour. To accomplish the landing, the X-15 used nose wheels, two skids on the tail, and a parachute in the tail to slow it down. These landings provided information later used to make safe space shuttle landings.

High-Tech Materials

Most airplanes are made of titanium and other metals. They cannot withstand the high temperatures associated with high-speed flight. The X-15 used a heat-resistant alloy called Inconel X, a combination of nickel and chrome. When the X-15 traveled at over 4,000 miles per hour, the temperature on the outside of the plane reached more than 1,200 degrees Fahrenheit. The friction of the air against the outside of the plane during high-speed flight caused this high temperature. Later, NASA put a chemical coating on the X-15, which allowed it to withstand temperatures over 2,000 degrees Fahrenheit.[3]

The X-15 was a dangerous airplane to fly. On November 5, 1959, pilot Scott Crossfield radioed that his X-15 displayed a fire warning light. A moment later there was an explosion in the engine. The X-15 lost power. It still had a full load of fuel, and it began falling like a stone. This was a serious situation. Crossfield had to try to land the aircraft. As a trained test pilot, he was cool under the pressure and concentrated intensely. When flying the X-15, a pilot had to remember his training and try to avoid a sudden reaction, such as bailing out.[4]

Crossfield had practiced this type of landing many times in an F-104 jet. The landing should have been routine. It was not. The engine explosion had weakened the aircraft's structure. When the nose wheels touched down, the airplane crumbled and almost broke in half. Amazingly, Crossfield was not hurt.

Major Michael J. Adams was not as fortunate. He was killed in an X-15 accident on November 15, 1967. The airplane broke apart during an uncontrolled inverted dive.

Amazing Records of the X-15

The X-15 set some amazing records. On July 17, 1962, Air Force major Robert H. White flew the X-15 to an altitude of 314,750 feet, or 59.6 miles above Earth. Since space begins at 50 miles, White became the first American pilot to qualify as an astronaut. The X-15 also became the first piloted American spacecraft.

On August 22, 1963, NASA test pilot Joseph A. Walker broke the altitude record for a piloted-winged aircraft. He flew the X-15 to 354,200 feet, more than 67 miles above Earth.

On October 3, 1967, Air Force test pilot Peter Knight flew the X-15 to Mach 6.7, or 4,520 miles per hour. The X-15's outside temperature reached 3,000 degrees Fahrenheit. It came within a few seconds of disintegrating.[5]

The X-15 set an unofficial altitude record of 354,200 feet, more than 67 miles above Earth. It also set an unofficial speed record of 4,520 miles per hour, or Mach 6.7.

The X-15's speed and altitude records for a winged aircraft remained unbroken until April 14, 1981, when the space shuttle *Columbia* circled Earth at about 17,500 miles per hour.

The X-15 led to an understanding of the heat effect on metals and humans at speeds above Mach 5. It helped scientists learn what kinds of materials were needed to build a space shuttle. The navigation equipment pioneered on the X-15 later helped guide the Apollo space capsules that first carried astronauts to the Moon.

The first *A* in NASA stands for *aeronautics*. Many of today's aircraft are the result of research by NASA. NASA continues its research by exploring space and incorporating developments in space, Earth science, and aerospace technology. One way it conducts research is by flying space shuttle missions.

CHAPTER 2

Riding the Space Shuttle

The space shuttle blasts from Earth's surface at Cape Canaveral, Florida. It carries astronauts and equipment into orbit around Earth. Once in orbit, astronauts can perform experiments to learn more about the planet or about microgravity. Other missions take supplies and astronauts to and from the International Space Station.

The space shuttle is a reusable spacecraft, about the size of a DC-9 commercial jetliner. However, its shape is different. Its wings begin just below the cockpit and extend back to the tail. Looking at it from above, it is shaped like a triangle. For launch, it is attached to an orange, 154-foot external liquid fuel tank. The fuel is a supercold mix of liquid oxygen and liquid hydrogen. The engines burn more than 1.8 million pounds (about 669,000 gallons) of this fuel at 1,000 gallons per second.[1] They produce the thrust needed to launch the shuttle.

In 2000, the space shuttle *Endeavour* completed a radar topography mission to map Earth's surface. This mission had a crew of six: a shuttle commander, a pilot, and four mission specialists.

≡ Launch Preparations

It is 8:00 A.M. on Friday, February 11, 2000. The space shuttle crew climbs the launch tower and boards *Endeavour*. They strap in and get ready for the launch. The hatch closes and the astronauts are now sealed in the spacecraft.

At T minus seven minutes, the orbiter access arm pulls back from the shuttle. The orbiter access arm is like a bridge between the shuttle and the launch tower.

At T minus six seconds, the main engines fire. Huge billowing columns of flame and smoke surround the base of the shuttle. Within four seconds, the engines are burning at 90 percent power.

At T minus zero seconds, the two white solid-fuel rocket boosters (SRBs) ignite. These are attached to each side of the huge external fuel tank. Each of the 149-foot-tall boosters carries more than one million pounds of solid fuel. They help the main engines lift the shuttle and accelerate it to more than 2,800 miles per hour.

12:43 P.M. Liftoff! This is *Endeavour*'s fourteenth mission.[2] Five miles away, hundreds of spectators can feel the ground beneath them shaking.

The shuttle shakes and roars as about 37 million horsepower of thrust pushes it off the launchpad. The shuttle is moving at 100 miles per hour by the time it clears the launch tower. The liftoff pushes the astronauts into their seats and they feel three times

Endeavour lifts off the launch pad at Cape Canaveral. The engines are burning more than 650,000 gallons of liquid oxygen and liquid hydrogen. The launch tower is seen to the left.

their body weight pressing down on them. Astronaut Roberta Bondar described the feeling: "Imagine lying on your back with a full-grown gorilla sitting on your chest."[3] The astronauts' heartbeats, which are normally around 85, rise to about 130 beats per minute.[4]

Two minutes after launch, the now empty SRBs fall away. Parachutes help them land gently in the ocean so that they can be reused on a future mission.

About eight minutes into the launch, the three main engines on the shuttle shut down. The empty external fuel tank is released. It burns up as it falls through the atmosphere.

The shuttle is now out of sight, about seventy miles above Earth. Only a long white contrail marks its flight path. It is traveling at approximately 17,500 miles per hour, about ten times faster than a bullet.

Once the external fuel tank and the two SRBs are gone, the small maneuvering rockets on the shuttle begin to place the spacecraft in orbit. These rockets are part of the orbital maneuvering system (OMS). A pod on the back of each side of the shuttle's body holds an OMS engine. The pod also has fourteen rockets and their fuel. There are sixteen more rockets in the nose of the shuttle. These rockets control the pitch. Others control the roll, keeping the wings level. Another set of rockets controls the side-to-side movement.

Free-Falling in Microgravity

Forty-five minutes after launch, the astronauts are in orbit. They are about 150 miles above Earth. The acceleration that pressed the astronauts into their seats is gone. They are free-falling in the microgravity of space. The shuttle is pressurized and provides the air and temperature the astronauts need to survive in space. The atmosphere in the cabin is like that inside an airliner.

Once the shuttle is in orbit, the payload bay can be opened. The payload bay is a large section toward the back of the shuttle, behind the flight deck. Along one side of the bay is a fifty-foot remote manipulator system called the robotic arm. It has a movable shoulder joint, elbow, and wrist. It also has a hand with four claws. It can grasp objects from orbit, such as satellites that need to be repaired.

For this mission, the shuttle carries two special radar antennae in its payload bay. The mission specialists operate these antennae. The antennae help create three-dimensional maps of more than 80 percent of Earth's surface. To do this, the pilot turns the shuttle upside down. The antennae in the open payload bay point toward Earth and send radar signals to Earth's surface. These signals bounce back to the antennae. The instruments aboard the shuttle make about one trillion measurements of Earth's surface. It takes special computer programs to process the data into

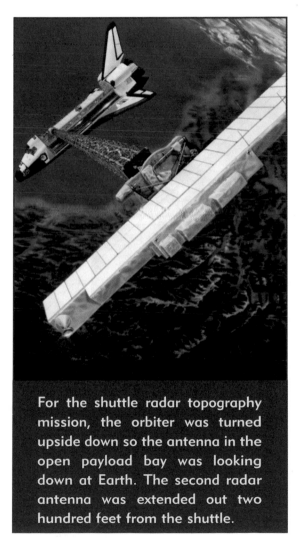

For the shuttle radar topography mission, the orbiter was turned upside down so the antenna in the open payload bay was looking down at Earth. The second radar antenna was extended out two hundred feet from the shuttle.

three-dimensional maps. The data will be used to produce global maps more accurate than any previously available.

≡ Going Home

On February 22, 2000, *Endeavour* receives permission to return to Earth. At 5:24 P.M., the commander and pilot direct the onboard computers to fire the small OMS rockets. These push the shuttle around to a tail-first position to slow it down. Then the space shuttle's nose is pointed upward, away from Earth. This way the 24,000 heat tiles on its belly will absorb the heat of reentry. As the shuttle descends through the atmosphere, the heat

NASA uses two modified Boeing 747s as Space Shuttle Carrier Aircraft (SCA). The shuttles normally take off and land at the Kennedy Space Center (KSC) in Cape Canaveral, Florida. If the shuttle has to land at Edwards Air Force Base in California, because of bad weather at Cape Canaveral, the SCA will fly the shuttle back to KSC for its next launch.

The Supersonic X-15 and High-Tech NASA Aircraft

tiles reach nearly 3,000 degrees Fahrenheit and glow orange-red.

At 400,000 feet, one of the five onboard computers moves the shuttle's wings to a level position and points the nose down. At 188,000 feet, the shuttle slows down to about 7,000 miles per hour.[5]

The space shuttle touches down after a successful mission.

About five minutes before touchdown, some residents near the Kennedy Space Center hear a sonic boom. *Endeavour* enters the atmosphere traveling at over 1,000 miles per hour.

At 49,000 feet, the shuttle slows to about 500 miles per hour. It is about 25 miles from the runway. At about 40,000 feet, the commander takes manual control of the craft and begins to line it up with the runway.[6]

At 6:22 P.M., *Endeavour* lands on a three-mile-long runway. Its touchdown speed is about 210 miles per hour. A parachute in the tail slows it down. The astronauts have completed a successful mission and a safe landing aboard the space shuttle.

CHAPTER 3

NASA's Jumbo Airplanes

The International Space Station (ISS) is the largest human-made object in space. Some parts of the ISS were built in the United States and other parts in Russia. Then they were carried into space aboard a space shuttle. Piece by piece, the station was assembled in space. Some assembly, including replacing worn parts, is ongoing.

Some of the pieces of the ISS are too large to fit on Earth-bound trains or trucks. They also cannot fit across bridges or through tunnels. NASA flies these parts to the launch site in a specially designed huge cargo airplane called the Super Guppy.

The Super Guppy is shaped like a whale.

It is a four-engine turboprop aircraft. The jet engines turn its propellers. The plane can cruise at about 290 miles per hour.

The Super Guppy has nine seats for the crew in the pressurized flight deck. The cargo bay, which is not pressurized, opens via the plane's enormous hinged nose. The nose opens to the side, and the bay can open wider than a barn door. In this way, huge pieces of cargo can be loaded and unloaded from the front.

Loading the Super Guppy is simple. The cargo is pulled on rails by an electric pulley beneath the cargo floor.

Although it is called a Super Guppy, it looks more like a whale. The Super Guppy's cargo compartment is 25 feet wide, 25 feet high, and 111 feet long. It can carry 52,500 pounds, more than 26 tons.

Special lock pins in each rail hold the cargo in place for flight.

≡ KC-135—The Vomit Comet

For training and test purposes, NASA uses several methods to create microgravity conditions on Earth. One way is by using a KC-135 aircraft nicknamed the Vomit Comet. This airplane is a Boeing 707, a former airliner.

The KC-135 flies in arcs. The pilot flies these arcs by pulling the nose up at a 45-degree angle until the plane will climb no higher. The top of the arc is reached at 36,000 feet, when the passengers experience about two gs. At the top of the arc, the pilot pushes the stick forward and continues to pitch over until the aircraft is 45 degrees nose down. This will cause the people inside to free-fall, becoming nearly weightless. During a flight, which lasts about two hours, the crew may fly forty arcs. This flight is called the vomit comet because sometimes the motion of the plane makes the astronauts sick.

The KC-135 climbs to 36,000 feet. Then it heads into a dive, producing microgravity inside.

The Boeing 747

Another of NASA's airplanes is the Boeing 747, a four-engine jumbo jet that can fly about 500 miles per hour. Ordinary wings of jet airplanes create invisible trails of dangerous spiraling air. If you could see them, they would look like twisting ropes in the sky. These are called vortexes. They cause wake turbulence, or rough air. Vortexes can make an airplane ride very bumpy. They can even make small aircraft go out of control. NASA has used the Boeing 747 to study these vortexes.

Since the vortexes are invisible, NASA scientists used smoke generators under the wings of the 747 to mark them. (Smoke generators are also used on airplanes at air shows.) The smoke generators created a visual image of the vortexes. NASA then put different devices on and

Two aircraft, a Learjet and a Cessna T-37, fly through the wake off the right wing tip of a Boeing 747 jetliner. The vortex trail behind the right wing tip of the 747 was made visible by a smoke generator mounted under the wing of the aircraft.

under the wings. They wanted to see which ones would break up or weaken the vortexes. Today small pieces of metal can be found on top of or underneath the wings of some jetliners. These reduce the vortexes and make air travel safer and more comfortable.

In 2005, another Boeing 747 will be used to carry a huge telescope that will help scientists study heat from stars. This study is called infrared astronomy. NASA will cut a huge hole into the side of a 747 and cover the hole with Plexiglas. This will create a giant window in the fuselage, or body of the plane. The scientists will put a telescope in the window and look into space.

This will be the largest airborne telescope in the world. It will enable scientists to see things that are impossible to see from ground-based telescopes. Scientists will study the heat patterns from stars and the makeup of comets.[1]

≡ Launching Satellites

NASA began launching satellites soon after it formed in 1958. Most often, it launches satellites from large rockets, but sometimes it launches them from airplanes in flight. In October 2000, NASA launched a satellite over the Pacific Ocean. It used a Lockheed L-1011 jumbo jet. The L-1011 was once a passenger airliner. It has three jet engines: one on each wing and one in the tail.

A Hybrid Pegasus rocket was mounted beneath the airplane. On the rocket was a 273-pound satellite called HETE-2 (High-Energy Transient Explorer). When the jet

NASA's L-1011 "Stargazer" is taking off with the X-34 reusable rocketplane underneath. The X-34 is capable of flying up to eight times the speed of sound.

reached 40,000 feet over the Pacific Ocean, NASA released the rocket. It fell for five seconds. Then the first-stage rocket engine fired. Ten minutes later, the rocket had placed the satellite into orbit. NASA controlled the launch from computers at Kennedy Space Center, Florida, more than halfway around the world.

HETE-2 monitors space for gamma ray bursts. These are the most powerful bursts of energy in the universe. When they occur, they last only a few seconds. Scientists will try to discover the source and location of this energy.[2]

While some of NASA's airplanes study space, others are studying Earth's weather, pollution, and geologic events such as earthquakes and volcanic activity.

CHAPTER 4

Earth Science Research

Many of the airplanes used for Earth science research fly from the Dryden Flight Research Center, at Edwards Air Force Base, in California. Dryden operates some of the most advanced research aircraft in the nation. These airplanes collect information about Earth's surroundings, including the planet's resources. Scientists also conduct experiments in the atmosphere. Some of these experiments include using high-altitude aircraft like the Lockheed ER-2 and the SR-71 for high-speed and high-altitude research.

Dryden airplanes collect information on weather, including air temperatures, and study ocean currents and ice movements. "We fly an airborne Earth science mission," said Mark Pestana, a pilot on the NASA DC-8. "We want to get a better understanding of what is happening to the Earth. We are using airplanes like the DC-8 and the ER-2 to discover the answers."[1]

"For some experiments," Pestana said, "we use radar. This radar makes images of the West Coast along the Cascade Mountains. Over time, we compare these images. We can tell if there has been volcano activity on Mount St. Helen's. We can also tell if there have been movements in the earthquake fault lines. If we can predict future volcano eruptions and earthquakes, we can warn people."[2]

NASA's Douglas DC-8 flies over one of the two Boeing 747s (on the ground) that carry the space shuttle.

≡ Antarctic Ozone Experiments

In 1987, the DC-8 and one of NASA's sleek ER-2s took part in a series of Antarctic ozone experiments. Ozone is a layer of air that blocks some of the Sun's dangerous rays. Recently, the ozone has been slowly disappearing. Scientists suspect that some human-made chemical compounds may be causing this. "We are discovering how human and natural causes combine to influence our air quality," said Pestana.[3]

The DC-8 performed seven experiments. Scientists sampled air through different openings, called optical ports, in the front of the plane.

The DC-8 looks like a regular airliner on the outside, but on the inside it is really a flying laboratory.

The NASA ER-2 was once a high-flying secret spy plane. It can fly higher than 85,000 feet.

One technique used aboard the DC-8 was called a "Sun run." The pilot flew the aircraft at right angles to the Sun. This enabled the Sun's rays to shine directly into the instruments. The instruments used sunlight to identify different chemicals in the atmosphere.

The DC-8 also uses synthetic aperture radar (SAR). Because of its wavelength characteristics, this radar can penetrate clouds, forest canopies, and soil. An SAR emits a microwave signal and uses the echo returned from leaves and Earth's surface.

"We can even 'see' below the soil a few inches," said Pestana. "From this we can tell how much moisture is in the soil. We can also tell if the plant life is healthy or not."

"A few years ago we were flying a mission over Cambodia," said Pestana. "We were looking for damage to the rain forests. Over one area, our synthetic aperture radar 'saw' through the jungle canopy. We discovered old religious temples that no one knew about."[4]

The ER-2 was also used in the Antarctic experiments. It can fly much higher than the DC-8. The ER-2 carried fourteen experiments that were installed in bays and pods located in the body of the plane and in both wings.

The ER-2 does not have standard landing gear. It has only one wheel on the bottom of the fuselage. For stabilization at takeoff, it has wheels on its wingtips called pogos. These drop from the wings as the ER-2 leaves the runway.

Hurricane Hunters

Hurricanes are dangerous and sometimes deadly. NASA goes hunting for them in order to find out how strong they will grow and where they might hit land. NASA has been using satellites to monitor hurricanes since the 1960s. Now it uses satellites, the ER-2, and the Douglas DC-8.

The DC-8 and its daring crew make several flights during a hurricane. They probe and measure the weather conditions ahead of the storm. They measure the hurricane's growth at low, medium, and high altitudes, since the moisture and energy throughout the storm will affect how strong the storm may become.

The DC-8 monitors the hurricane's wind speeds, barometric pressure, and eye location. One of the maneuvers the DC-8 performs is a spiraling turn down from about 35,000 feet to 14,000 feet to sample the rain. During these flights, the DC-8 uses instruments to measure the storm's structure, environment, and changes in intensity.

While the DC-8 is plunging through the storm, the specially equipped ER-2 flies above the storm at 65,000 feet. It measures the surrounding atmosphere that steers the storm's movement. This way NASA scientists can predict the hurricane's movement and strength when it first lands. This enables them to warn people and save lives before the hurricane strikes.

The DC-8 monitors hurricane wind speeds and barometric pressure. This plane can fly up to 600 miles per hour and 41,000 feet high.

The ER-2 can fly so high, it has even collected meteor dust. Scientists are studying the dust, hoping to learn more about the universe. "The ER-2 is providing a new way to understand the changes on Earth that we humans create," said Pestana. "We can then tell how they will affect life."[5]

Helios for Power

NASA has been working for years on developing nonpolluting energy. One source of nonpolluting energy is solar power. Solar power is energy created from the light of the Sun. One aircraft that uses solar power is the Helios Prototype.

The solar-powered Helios Prototype has fourteen electric motors to turn the fourteen propellers. The motors get their power from the solar cells on the wings.

The Helios Prototype is an unpiloted aerial vehicle. It is being developed for long flights at high altitudes to sample the atmosphere. Although it is only 12 feet long, the Helios Prototype has a wingspan of 247 feet—wider than a Boeing 747 or Lockheed C-5 transport aircraft, and almost the length of a football field. There are approximately 65,000 solar cells across the wings. The energy from the cells drives fourteen electric motors that turn fourteen propellers, each 79 inches in diameter. Combined, these solar cells can produce more than 35,000 watts of electricity. (An average household lightbulb produces about 75 to 100 watts.)

On August 14, 2001, the Helios Prototype solar-powered flying wing set an altitude record of more than 96,500 feet. No other non-rocket powered aircraft in horizontal flight has reached this height.[6]

In the near future, the Helios could permit long-term monitoring of climate patterns, changes in vegetation, and other environmental developments. It could also serve as a telecommunications relay station in places where orbital satellites are not available. It could orbit over cities at 55,000 to 70,000 feet above the weather, for months at a time. It would be the equivalent of an 11-mile-high radio and television tower in the sky.

NASA airplanes fly research missions from the Antarctic to space. They also do flight research in the laboratory using wind tunnels and supercomputers.

CHAPTER 5

Ground-Based Flight Research

When humans were first trying to build airplanes, they studied birds to learn how they flew. They soon realized that it would take more than watching birds to unlock the secrets of flight. For people to fly, they needed to understand the flow of air over moving surfaces and how to use this information to improve aircraft design. The Wright brothers, who made the first controlled powered flight, built wind tunnels to test their designs. Soon, aircraft builders realized that wind tunnels could test the wings and fuselages under safe and controlled conditions. Today every new aircraft design undergoes wind-tunnel testing before being built to full size and test-flown.

Wind tunnels measure the flow of air over aircraft, spacecraft, and rocket surfaces. They also predict the craft's flight characteristics. They create the same conditions an aircraft would encounter if it were flying through the atmosphere. Wind tunnels allow researchers to safely take measurements that are often impossible to take while an aircraft is in flight.

A wind tunnel works by having a fan drive air through the tunnel at high speeds. The air passes over a scale model of an aircraft or wing suspended in the tunnel. Computers record how the object moves at different wind speeds. NASA has wind tunnels that can simulate speeds ranging from zero to nearly 17,500 miles per hour.[1] The fastest is the speed of the space shuttle. NASA built a one-third-sized model of the space shuttle and tested it in the wind tunnel before the shuttle was ever flown.

Supercomputers

Supercomputers are extremely powerful computers. The term is commonly used for the fastest high-performance computer systems available. Before the age of computers, engineers designed airplanes using complex mathematical formulas. They calculated results on manual slide rules. This was a very slow process and led to mistakes.

Today, scientists use supercomputers to solve complex aerodynamic equations. They even use computers to simulate aircraft in flight and safely test aircraft designs. Supercomputers provide scientists with several possible

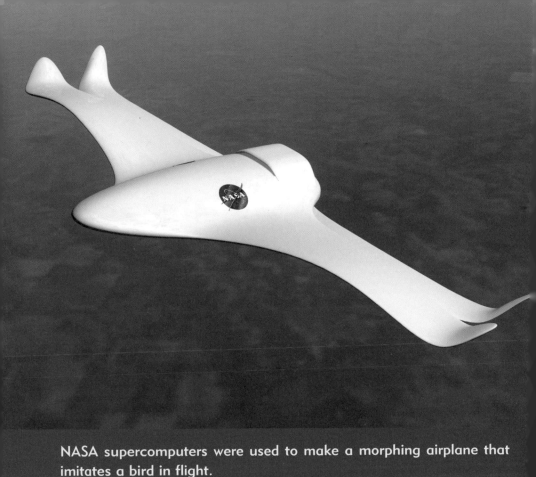

NASA supercomputers were used to make a morphing airplane that imitates a bird in flight.

solutions to the same aircraft design problem. Some NASA supercomputers are able to compute one trillion (1,000,000,000,000) operations per second.[2]

Computers are helping NASA develop a morphing (changing) airplane that would imitate a bird in flight. Sensors in the skin and structure could detect variations in pressure and other flight characteristics. Computers would drive mechanical muscles, bones, and skin, adjusting the airframe and wings as a bird does to maintain its best possible flight shape.

≡ Blended Wing Body

One airplane design being tested in both the wind tunnel and on the computer is the blended wing body (BWB). Air travelers of the future may step onto a double-deck jetliner that resembles a flying wing. NASA researchers estimate a BWB could carry up to 500 passengers. They predict that it could fly for 7,000 miles at a speed of about 560 miles per hour.

Wind-tunnel and computer testing are necessary because the revolutionary BWB design has a thick, wing-shaped fuselage. Today's aircraft fly at speeds approaching 600 miles per hour. At these speeds, the thick wing of the BWB would experience high aerodynamic drag. Current airliners with their thin wings do not experience this problem as severely. The wind tunnel will simulate the airflow at low and high speeds. Wind-tunnel testing will help determine the performance and the stability of the BWB's design.

The scale model of the blended wing body has two passenger decks and three engines.

Cabin pressure is another of the design challenges. Current airliners have a cigar-shaped fuselage ideal for maintaining cabin pressure. The BWB, however, has a unique shape that requires a new approach to pressurization needs. Computers can simulate different cabin pressures, and engineers can modify the fuselage design in the computer until they get satisfactory results.

The numerous types of NASA aircraft fly many different missions. They fly in the atmosphere like airliners, and one even flies into space. They carry cargo and fly Earth science missions to help identify pollutants that are damaging the air on Earth. Some are used for the development of space and aerospace technology.

A scale model of the blended wing body hangs in a wind tunnel. The wind tunnel can measure the flow of air over the aircraft. Researchers can safely measure the wind's effects before the aircraft takes flight.

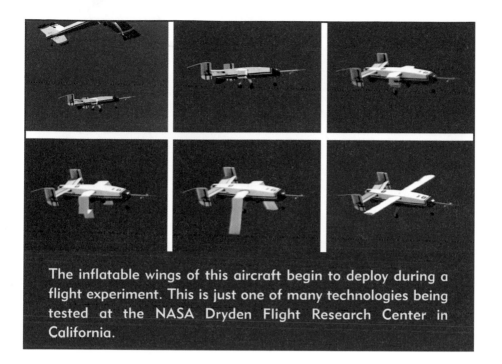

The inflatable wings of this aircraft begin to deploy during a flight experiment. This is just one of many technologies being tested at the NASA Dryden Flight Research Center in California.

"NASA is about more than space—we're about life on Earth . . . ," said former NASA Administrator Daniel S. Goldin. "Our Earth Science program is . . . providing down-to-Earth benefits. For instance, our data helps improve agricultural management by identifying disease susceptibility, assessing soil moisture, and helping farmers determine how much fertilizer to use where."

"Future Earth Science research will help us better understand our own planet."[3]

Specifications for
NASA Aircraft

Boeing B-52

Length—157 feet 7 inches

Height—40 feet 8 inches

Wingspan—185 feet

Engines—8

Top speed—About 650 miles per hour

Maximum altitude—About 50,000 feet

Range—8,800 miles

Boeing B-52

North American X-15

Length—50 feet 7 inches

Height—14 feet

Wingspan—22 feet 4 inches

Engines—1

Top speed—4,520 miles per hour

Maximum altitude—354,000 feet

North American X-15

Space Shuttle

Orbiter

Length—122 feet

Height—57 feet

Payload bay—60 feet long and 15 feet wide

Wingspan—78 feet

External Tank

Length—154 feet

Diameter—28 feet
7 inches

Solid Rocket Boosters—2

Height—149 feet
2 inches

Diameter—12 feet 2 inches

Space Shuttle and Launch Tower

Super Guppy

Length—143 feet 10 inches

Height—48 feet 6 inches

Wingspan—156 feet
3 inches

Engines—4

Cruise speed—290
miles per hour at
25,000 feet

Maximum altitude—About 45,000 feet

Maximum load—52,500 pounds

Super Guppy

Boeing 747

Length—231 feet 10 inches

Height—63 feet 5 inches

Wingspan—195 feet 8 inches

Engines—4

Cruise speed—About 500
  miles per hour

Maximum altitude—About 45,000 feet

Boeing 747

Douglas DC-8

Length—157 feet

Wingspan—142 feet 5 inches

Engines—4

Top speed—About 600 miles
  per hour

Maximum altitude—About
  41,000 feet

Douglas DC-8

Lockheed ER-2

Length—62 feet 1 inch

Height—16 feet

Wingspan—103 feet 4 inches

Engines—1

Cruise speed—More than
  475 miles per hour

Maximum altitude—Above 85,000 feet

Range—2,500 miles

Lockheed ER-2

Chapter Notes

Chapter 1. Flying the Supersonic X-15

1. Richard Tregaskis, *X-15 Diary—The Story of America's First Space Ship* (New York: Dutton, 1961), p. 227.

2. William F. Dana, "From the Pilot's Seat," *Science News*, February 24, 1968, <http://jsbim.sourceforge.net/X-15.html#Dana> (October 23, 2001).

3. Jeffrey L. Ethell, "At the Threshold of Space," *Air & Space Smithsonian*, October/November 1993, p. 27.

4. Tregaskis, p. 215.

5. Ethell, p. 33.

Chapter 2. Riding the Space Shuttle

1. Dryden Flight Research Center, *NASA Facts*, "Space Shuttles," <http://trc.dfrc.nasa.gov/PAO/PAIS/HTML/FS-015-DFRC.html> (October 23, 2001).

2. NASA, *Kennedy Space Center*, "STS-99(97)," <http://science.ksc.nasa.gov/shuttle/missions/sts-99/mission-sts-99.html> (October 23, 2001).

3. Roberta Bondar, *On the Shuttle* (New York: Firefly Books, 1993), p. 10.

4. Andrew Wilson, *Space Shuttle Story* (New York: Crescent Books, 1986), p. 59.

5. Tim Furniss, *Space Shuttle Log* (London: Jane's, 1986), p. 36.

6. Debbie Gary, "Space Shuttle Impersonator," *Air & Space Smithsonian*, October/November 2000, p. 41.

Chapter 3. NASA's Jumbo Airplanes

1. NASA, *Stratospheric Observatory for Infrared Astronomy*, "Mission Schedule," <http://sofia.arc.nasa.gov/Sofia/status/sofia_status.html> (October 23, 2001).

2. Craig Covault, "Kennedy Manages Pacific Launch," *Aviation Week and Space Technology*, October 16, 2000, pp. 61–62.

Chapter 4. Earth Science Research

1. Author interview with Mark Pestana, July 30, 2000, Oshkosh, Wis.

2. Doug Stewart, "Above the Sky," *Air & Space Smithsonian*, August/September 1993, p. 27.

3. Author interview with Mark Pestana, July 30, 2000, *AirVenture 2000*, Oshkosh, Wis.

4. Ibid.

5. Ibid.

6. NASA, *Dryden Flight Research Center*, "Helios," November 2, 2001, <http://www.dfrc.nasa.gov/Projects/Erast/helios.html> (November 23, 2001).

Chapter 5. Ground-Based Flight Research

1. Frederick A. Johnsen, "NASA Solar Airplane Soars on Sunshine," *AirVenture Today*, July 25, 2001, p. 15.

2. NASA Facts FS-1999-10-20-LaRC, October 1999.

3. NASA, "NASA Administrator Goldin's Oral Testimony on NASA FY2000 Posture," February 24, 1999, <http://www.hq.nasa.gov/office/ese/nra/gspeech/index.html> (October 23, 2001).

Glossary

aerodynamic drag—A force opposing an aircraft's forward motion caused by friction with the air.

air-launch—To carry an aircraft, rocket, or space vehicle aloft underneath another aircraft and release it.

alloy—A combination of two or more metals.

anhydrous—A substance that does not contain any water.

contrail—Streaks of condensed water vapor formed by hot engine exhaust in the subzero temperatures at high altitude.

control surface—A movable airfoil designed to modify the attitude or direction of an aircraft.

fuselage—The central portion, or body, of an aircraft that holds the crew, passengers, and/or cargo.

g-force—The force exerted on a person or object caused by a change in acceleration.

gravity—A force between two or more objects that causes them to be attracted to each other.

International Space Station—A permanent laboratory orbiting Earth that will enable long-term research in space. It is the largest international scientific project ever attempted.

lift—The force on a wing created by airflow.

Mach number—A number used to compare very high speeds. Mach 1 equals the speed of sound, which is about 750 miles per hour at sea level. Mach 3 is three times that speed.

microgravity—The almost complete absence of gravity.

orbit—The path of an object in space around another object in space.

payload—The load, or cargo, carried by an aircraft or spacecraft.

pitch—To point up or down.

roll—The sideways rotation of an aircraft caused when one wing rises higher than the other.

skid—Part of the landing gear of an aircraft.

solid rocket boosters—The two rockets attached to opposite sides of the external fuel tank. They provide additional power for liftoff of the space shuttle.

sonic boom—A sound similar to an explosion. It is produced when a shock wave from the nose of a supersonic aircraft reaches the ground.

space shuttle—The first reusable spacecraft. It is made up of three main parts: the external tank, the orbiter, and the solid rocket boosters.

thrust—To push or drive with force.

topography—The features of a surface, such as Earth's; also, the mapping of these features.

turbulence—Irregular disturbances in the flight of an aircraft created by changes in air flow.

vortex—Invisible trails of dangerous spiraling air formed behind the wings of a jet plane as it flies.

Further Reading

Books

Ayres, Carter M. *Pilots and Aviation*. Minneapolis, Minn: Lerner, 1990.

Bilstein, Roger E. *Flight in America: From the Wrights to the Astronauts*. Baltimore: Johns Hopkins University Press, 1994.

Bond, Peter. *Guide to Space*. New York: Dorling Kindersley, Inc., 1999.

Cole, Michael D. *Astronauts Training for Space*. Berkeley Heights, N.J.: Enslow Publishers, Inc., 2000.

————. *NASA Space Vehicles: Capsules, Shuttles, and Space Stations*. Springfield, N.J.: Enslow Publishers, Inc., 2000.

Stott, Carole. *Space Exploration*. New York: Alfred A. Knopf, 1997.

Yount, Lisa. *Women Aviators*. New York: Facts on File, 1995.

Internet Addresses

Kennedy Space Center
   <http://www.ksc.nasa.gov>

NASA Homepage
   <http://www.nasa.gov>

NASA's Jet Propulsion Laboratory
   <http://www.jpl.nasa.gov>

NASA's Kids Corner
   <http://vraptor.jpl.nasa.gov/voyager/kids.html>

NASA Spacelink
   <http://search.spacelink.nasa.gov>